FORGOTTEN BEASTS

AMAZING CREATURES THAT
ONCE ROAMED THE EARTH

MATT SEWELL

PAVILION

For Romy & Mae, Eve & Fin and Arlo

CONTENTS

INTRODUCTION

Welcome to the amazing world of forgotten beasts!

You have heard of dinosaurs, the monsters that ruled the planet more than 65 million years ago. But do you know anything about the *Paraceratherium*, the rhino that was more than 16 feet 5 inches tall? Or the terror birds, a South American family of giant flesh-eating birds? How about pygmy mammoths, the tiny mammoths of California?

These weird and wonderful creatures, and many more, once called our planet home. Come inside to explore their stories—the beasts that time forgot.

A note on the illustrations

We have always imagined the beasts of our past to have been muddy brown or boring green. No longer! Some smart palaeontologists (these are scientists who study fossils) now think that many creatures may have been colorful. Penguins were once decked in trendy red, marsupials clad in tiger stripes, and turtles may have been pink.

The illustrations on the following pages have been inspired by these ideas— helping you to really imagine the bright, scary world of the distant past.

A scale of time

We begin our journey with the *Opabinia*, a bizarre creature from the Cambrian period more than 500 million years ago, and we end it with the *Thylacine*, the last of which walked the earth around 80 years ago. How can we even begin to comprehend such scale?

Below you can see a timeline.* It will help you to understand the scientific names of the periods along with the beasts that lived in those times. The beasts in this book have been arranged chronologically, so the more you read, the closer you will get to the present day.

Diets

Such a wide range of magnificent creatures ate a wide range of food. The *Odobenocetops* fed on mollusks buried in sediment, the *Paraceratherium* would have browsed on leaves and the *Smilodon* ate large mammals, such as bison and giant sloths.

Beast diets can be described using these terms:

Carnivorous—meat eating
Herbivorous—plant eating
Omnivorous—meat and plant eating
Piscivorous—fish eating
Insectivorous—insect eating
Detritivorous—detritus (waste) eating

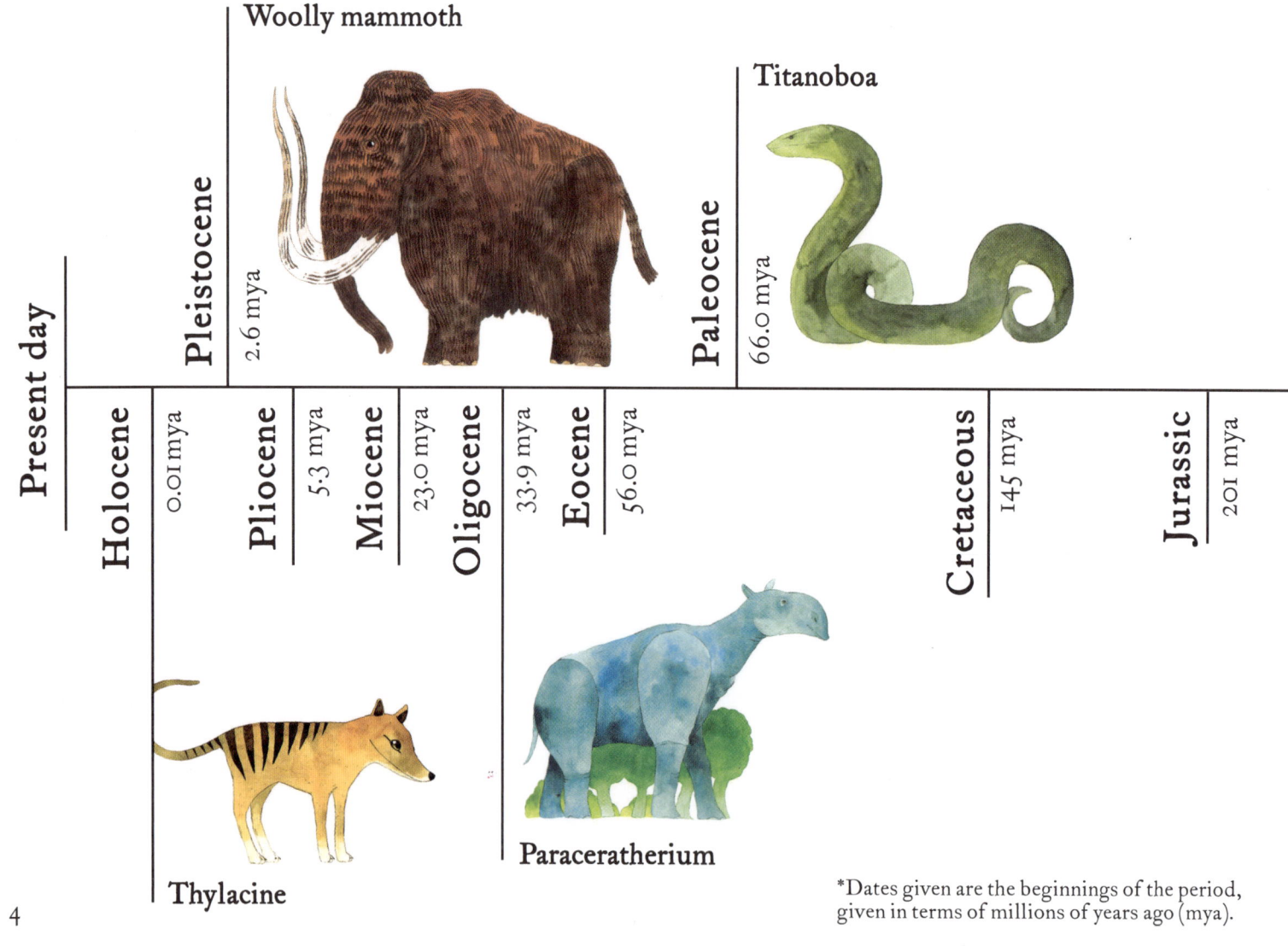

Present day

Holocene — 0.01 mya

Pleistocene — 2.6 mya

Woolly mammoth

Pliocene — 5.3 mya

Miocene — 23.0 mya

Oligocene — 33.9 mya

Eocene — 56.0 mya

Paleocene — 66.0 mya

Titanoboa

Cretaceous — 145 mya

Jurassic — 201 mya

Paraceratherium

Thylacine

*Dates given are the beginnings of the period, given in terms of millions of years ago (mya).

Forgotten beasts . . . of the future?

As you turn the pages of this book, you will start to notice patterns in the causes and reasons behind each extinction. For many of the beasts, irreversible climate change was the deciding factor. For others, clashes with humans assigned their fate.

There are still many wonderful creatures that live wild all over our planet. We must do everything in our power to be sure that they do not become forgotten beasts, too.

DID YOU KNOW?

The dire wolf did not target easy kills, such as elk and deer, but preferred to hunt fast-running horses and imposing bison instead.

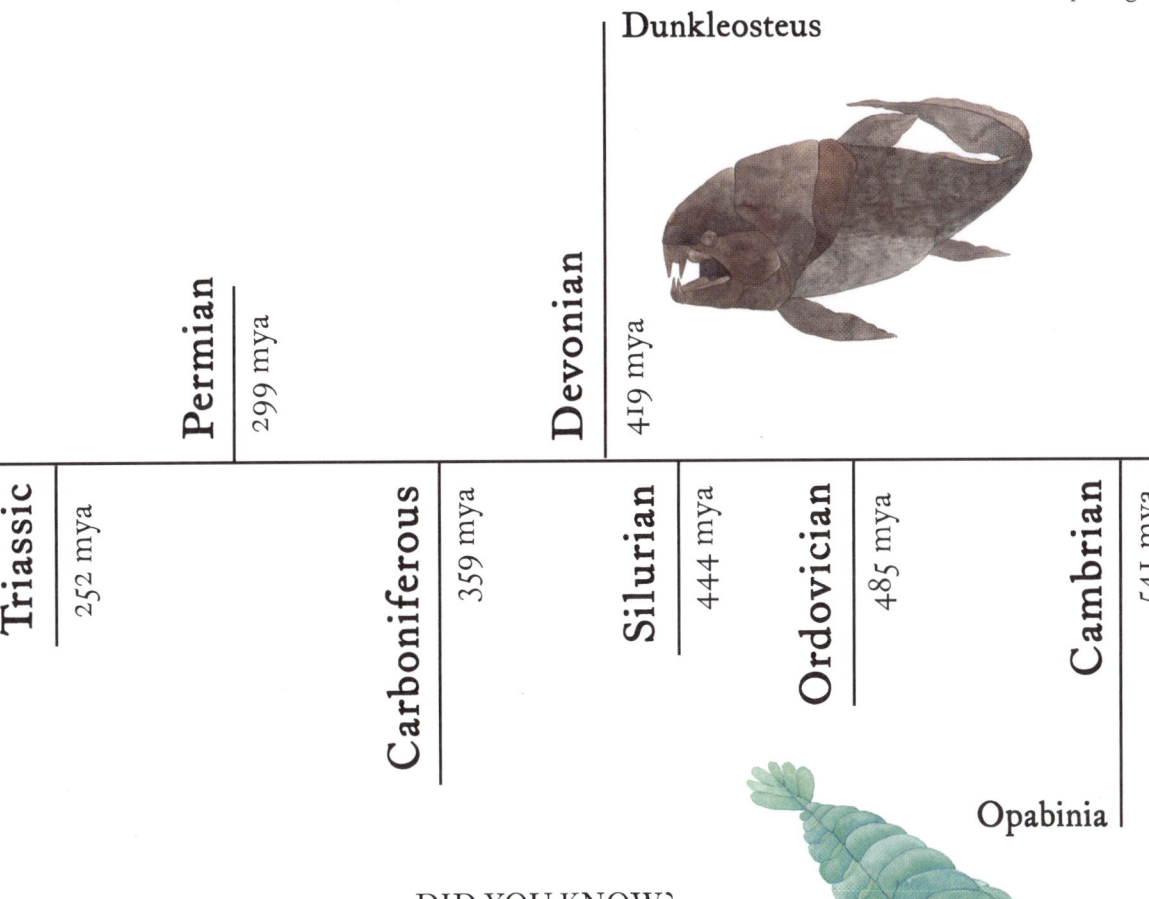

Dunkleosteus

Triassic	Permian	Carboniferous	Devonian	Silurian	Ordovician	Cambrian
252 mya	299 mya	359 mya	419 mya	444 mya	485 mya	541 mya

Opabinia

Now the time has come to bring these monsters back to life.

Are you brave enough to turn the page?

DID YOU KNOW?

The *Opabinia* lived in an age where there were two continents, Gondwana and Laurentia, and all life was found in the sea—including plants. Compare that to the *Thylacine*, which was still alive at the beginnings of photography and the automobile.

Opabinia

Size: 2 inches long
Weight: 5 ¼ ounces
Time period: Mid-Cambrian
Diet: Carnivorous and detritivorous

This sea oddity from 508 million years ago has been found in western Canadian shale. It is an arthropod, meaning an invertebrate with an exoskeleton, but it is not like any other creature, alive or dead. The *Opabinia* was only 2 inches long and was propelled by its side flaps and fanned tail. Its head was crowned by five eyes on stalks and its mouth was out of sight underneath. If that is not weird enough, it also had a clawlike proboscis on the end of a ribbed tube, which was used for catching food and placing it in its mouth. Deep-sea research is heralding new species every year, but nothing has been discovered as odd as this weirdo.

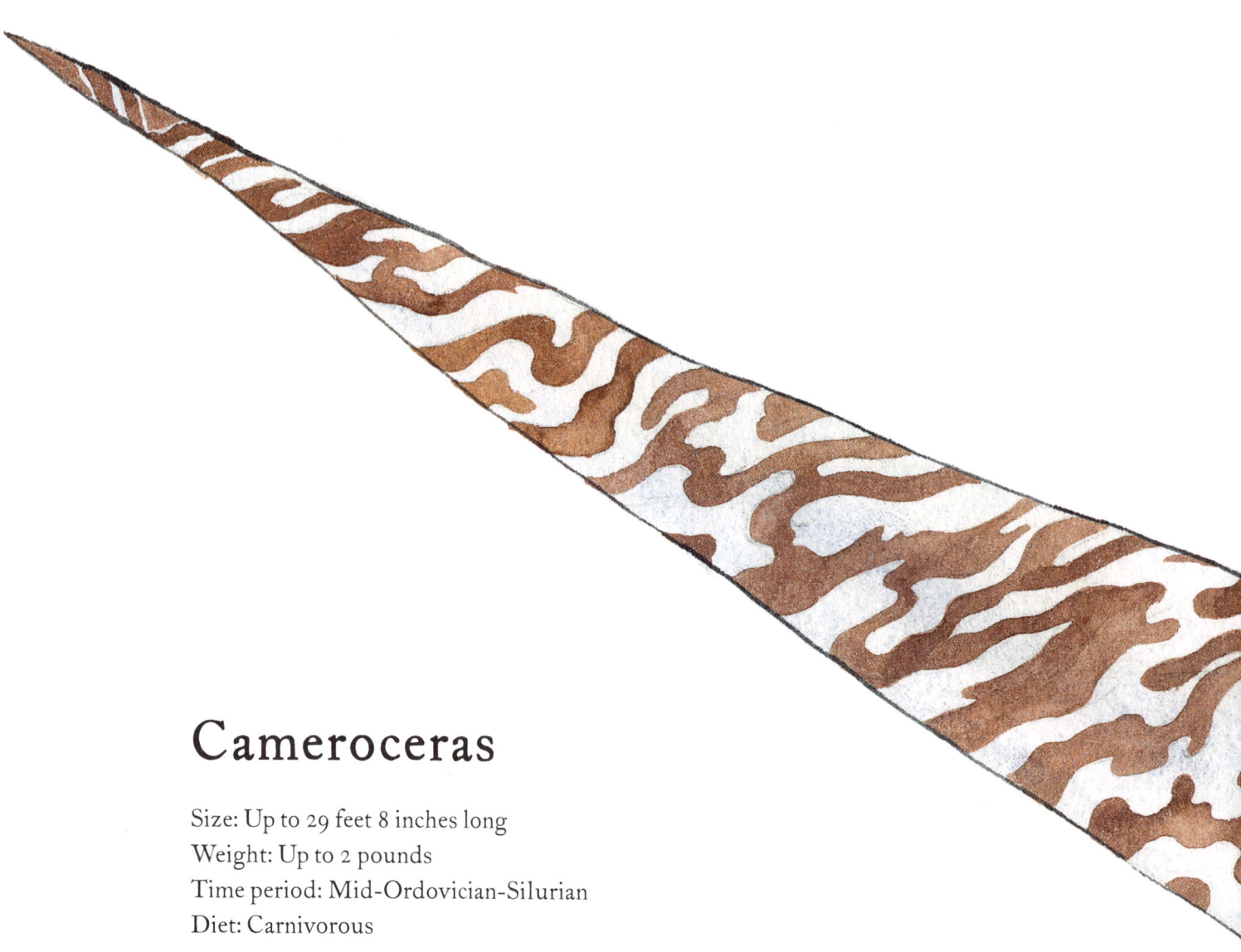

Cameroceras

Size: Up to 29 feet 8 inches long
Weight: Up to 2 pounds
Time period: Mid-Ordovician-Silurian
Diet: Carnivorous

The conical *Cameroceras* would have been a regular sight in the lively Ordovician seas of more than 450 million years ago. This giant orthoconic cephalopod (predatory marine mollusk) was a distant relative of squid and cuttlefish. It would have propelled itself through the water, dragging its shell with it, much like today's beautiful chambered nautilus, another relative.

The 29-foot 8-inch-long shell would have provided shelter from even bigger and meaner underwater beasts that lurked in the deep. It also would have been a place of danger for many fish and mollusks, because behind that tangle of tentacles was a sharp beak that could pierce the toughest of shells.

Jaekelopterus

Size: 8 feet 3 inches long
Weight: Up to 620 pounds
Time period: Early Devonian
Diet: Carnivorous

Despite being a kind of sea scorpion, the
Jaekelopterus lived in freshwater rivers and lakes.
There it would lie in wait to snap and grab passing
fish, with its huge claws that were bigger than this
book. The *Jaekelopterus* was the biggest arthropod
that has ever lived, completely overshadowing
the creepy-crawlies we have today. Nothing comes
close, not even the large Japanese spider crab.

Dunkleosteus

Size: Up to 20 feet long
Weight: Up to 1 ton
Time period: Late Devonian
Diet: Piscivorous

The *Dunkleosteus* is a 360-million-year-old ancient beast from the deep. It was a huge, bizarre fish that had an unwieldy bone armor-plated body, ready for battle like a rusty warship. Its helmeted head was also similarly armored, and it had sharp teeth powerful enough to crack any shell.

Edestus

Size: 20 feet long
Weight: 2,200 pounds to 4,400 pounds
Time period: Late Devonian to Late Carboniferous
Diet: Piscivorous

More creatures from the ancient deep—this period was such a time of great tectonic shift that marine fauna ruled the earth. The Devonian is even known as the "Age of the Fish." We have discovered only a fraction of the life from that epoch, but there were many strange and wonderful things, such as these two sharks.

First the *Edestus*—closely related to the *Helicoprion* (see page 16) and just as odd. It had two sets of deadly teeth that would slash, shear, and slice like an ancient pair of scissors. Next is the *Stethacanthus*, which has what can only be described as an anvil on its back, or a little table, maybe. Who knows.

Stethacanthus

Size: 1 foot 7½ inches to 3 feet 3 inches long
Weight: 10 pounds to 20 pounds
Time period: Late Devonian to Early Carboniferous
Diet: Piscivorous

Helicoprion

Size: Up to 40 feet long
Weight: Up to 1 ton
Time period: Early to Mid-Permian
Diet: Piscivorous

The fossilized remains of the *Helicoprion* have long been puzzling to palaeontologists, because the clutter of bones always contained a strange whorl of teeth. Since the late 1800s, this huge fish has been described in all kinds of odd and unwieldy getups, with spiraling teeth coming from its fin, dragging from its tail, curling up from the snout, and many other ways. In fact, there are so many theories, it is hard to keep up, but the most common thinking is that the whorl of teeth was held in place like a regular jaw. It is almost like this ancient shark dropped and lost his false teeth and used a saw blade instead. Which is, funnily enough, what his name means—spiral saw.

Longisquama

Size: 6 inches long
Weight: 3 ½ ounces
Time period: Mid to Late Triassic
Diet: Insectivorous

The enigmatic *Longisquama* was a genus of reptile from the Triassic period, the early days of the dinosaurs, around 230 million years ago. Only one skeleton has been discovered, in a Kyrgyzstani rock formation, so this small curiosity has sent minds boggling over what the creature actually looked like. Were those strange protruding feather-like spines a fan, a mating display, a predator warning . . . or tiny golf clubs? Who knows, but maybe not the last one.

Sarcosuchus

Size: 40 feet long
Weight: 8³/₄ tons
Time period: Early Cretaceous
Diet: Carnivorous

If you think that saltwater crocodiles are scary (understandably, too), then you will really be worried about this one. The *Sarcosuchus* was twice the size of that already monstrously big beast. This brute was 40 feet long—the same length as a bus. To add to this vision of nightmares, it had a grotesque nub called a bulla at the entrance of its violent jaws. The purpose of the bulla is unknown, but some think it was to make a hideous sound. I am even more scared! The *Sarcosuchus* was around during the Cretaceous period, the same time as many dinosaurs, but this croc became extinct around 110 million years ago. Such a shame. Honest. No, really it was.

Titanoboa

Size: 46 feet long
Weight: 1 1/4 tons
Time period: Paleocene
Diet: Carnivorous

Paleocene South America was wall-to-wall with alpha predators, and this jungle-dwelling titanic boa was one of them. This boa constrictor granddaddy would have had a girth of more than 3 feet, and a length of about 46 feet, so coiled up it would have been big enough to fill a house. It would not just have suffocated its victims but completely crushed them, and it was capable of swallowing a full-grown crocodile whole. Spending most of its time under water to support its mass, the *Titanoboa* eventually succumbed to climate change. The earth was not a hospitable place for megasize monsters when it started getting too chilly.

Gastornis

Size: 6 feet 8 inches tall
Weight: 375 pounds
Time period: Paleocene to Eocene
Diet: Herbivorous

The *Gastornis* was another prehistoric anomaly. Remains of this widespread bird have been found in France, England, and even China and the United States, but little concrete information is known except that it was 6 feet 8 inches tall and had a massive skull. Although it looked like a killing machine and was big enough to catch small horses, it is actually believed that the *Gastornis* was as vegetarian as a goose. Maybe this is why it died out—it would have taken a lot of greens to keep this big bird going.

Rodhocetus

Size: 8 feet 3 1/2 inches long
Weight: 1,300 pounds
Time period: Early to Mid Eocene
Diet: Piscivorous

This ferocious whale might just be the perfect evolutionary illustration of how animals went from land to sea. From more than 40 million years ago, it is almost caught in the middle between sea creatures and the wolflike hippo predators of the time. It had legs with paws that were webbed, a mammal-like snout and teeth, and a long tail for steering.

Andrewsarchus mongoliensis

Size: 10 feet 8 inches long, 6 feet 8 inches tall
Weight: 1,765 pounds
Time period: Eocene
Diet: Carnivorous

I wouldn't get too friendly with the *Andrewsarchus mongoliensis*
(or Andy to his friends). This huge pig-hippo-whale relative was
the biggest meat-eating mammal that has ever walked the earth.
Known only by a 45 million-year-old megaskull from Mongolia,
scientists have estimated that this beast was 6 feet 8 inches tall,
10 feet 8 inches long, and of that length, about 10 feet would have
been its bone-smashing skull. Wow!

Inkayacu paracasensis

Size: 5 feet tall
Weight: 110 pounds to 130 pounds
Time period: Late Eocene
Diet: Piscivorous

Discovered in Peru in 2008, the *Inkayacu*, or "water king" to me and you, was a long-beaked penguin of the Eocene period. It stood at more 5 feet tall, 1 foot taller than today's Emperor penguins, and would have been a formidable ocean hunter. Exquisite 30 million-year-old examples of the bird's plumage have been found, which show gray and reddish tints instead of the blacks and whites of the penguins we now know. So it is not hard to imagine the water king in a slick and colorful wet suit instead of a formal tuxedo. (Not that it holds our penguins back—they are perfectly designed for the chilly Antarctic seas!)

Megacerops

Size: 100 feet long, 8 feet 3 inches tall
Weight: 3 ⅓ tons
Time period: Late Eocene
Diet: Herbivorous

This oddball is from the long and varied evolutionary family that includes the horse, the rhinoceros, and the tapir. It had a stout body and thick protective armor. But is that even a horn on its head? It is not sharp, long, or even scary looking. In fact, it looks more useful than threatening. Was it a handy implement for twisting the tops off jars? Maybe a tuning fork for folk songs from the North American Eocene? Or was it used as a catapult? No—in actuality, it would have been used for protecting baby *Megacerops*, and for fighting other *Megacerops* during the mating season.

Pelagornis sandersi

Size: 21-foot wingspan
Weight: 65 pounds
Time period: Oligocene to Pleistocene
Diet: Piscivorous

Some scientists have thought that the *Pelagornis* should have been too big and heavy to fly. This creature should have been riding waves like a toothed* and flightless seabird. The *Argentavis magnificens* (see page 48) was long thought to be the world's biggest ever bird, but it was shunted into second place when workers discovered a more enormous set of bones (belonging to the *Pelagornis*) while working on a South Carolinian airport in the 1980s. Those bones belonged to a cliff-dwelling seabird that sported a wingspan of almost 23 feet. For comparison, the wandering albatross, our current living record holder, has a wingspan of only 11 feet 8 inches. It is no wonder that scientists find it hard to see this seaworthy beast capable of flight.

* These teeth were actually more like sharp spikes designed to hold slippery fish.

Paraceratherium

Size: 24 feet 3 inches long, 16 feet 5 inches tall
Weight: 16½ tons to 22 tons
Time period: Oligocene
Diet: Herbivorous

The *Paraceratherium* was one of the biggest mammals ever, and what a whopper it was. This 26-foot 5-inch-tall hornless rhino would have towered above the modern-day rhinoceros, which still reaches a formidable 6 feet. The *Paraceratherium* could have relied on its sheer size for self-defense, instead of needing to charge at would-be predators. Only its young would have been seen as prey, so it probably lived in family herds, just like modern-day elephants, protecting the fledglings from all the other strange and wonderful beasts of the Oligocene epoch.

Phorusrhacos

Size: 8 feet 3 inches tall
Weight: 285 pounds
Time period: Early to Mid-Miocene
Diet: Carnivorous

The Phorusrhacidae family of the Paleocene to
Pleistocene periods goes by a more memorable name—
the terror birds. There were a few species in this group,
but they all had a few things in common—they were
large, flightless terrifying carnivores. The 8-foot
3-inch-tall *Phorusrhacos* was one of them. It was believed
to have taken up the mantle of the region's apex predator
after the demise of the dinosaurs, a title held proud until
its disappearance 1.8 million years ago.

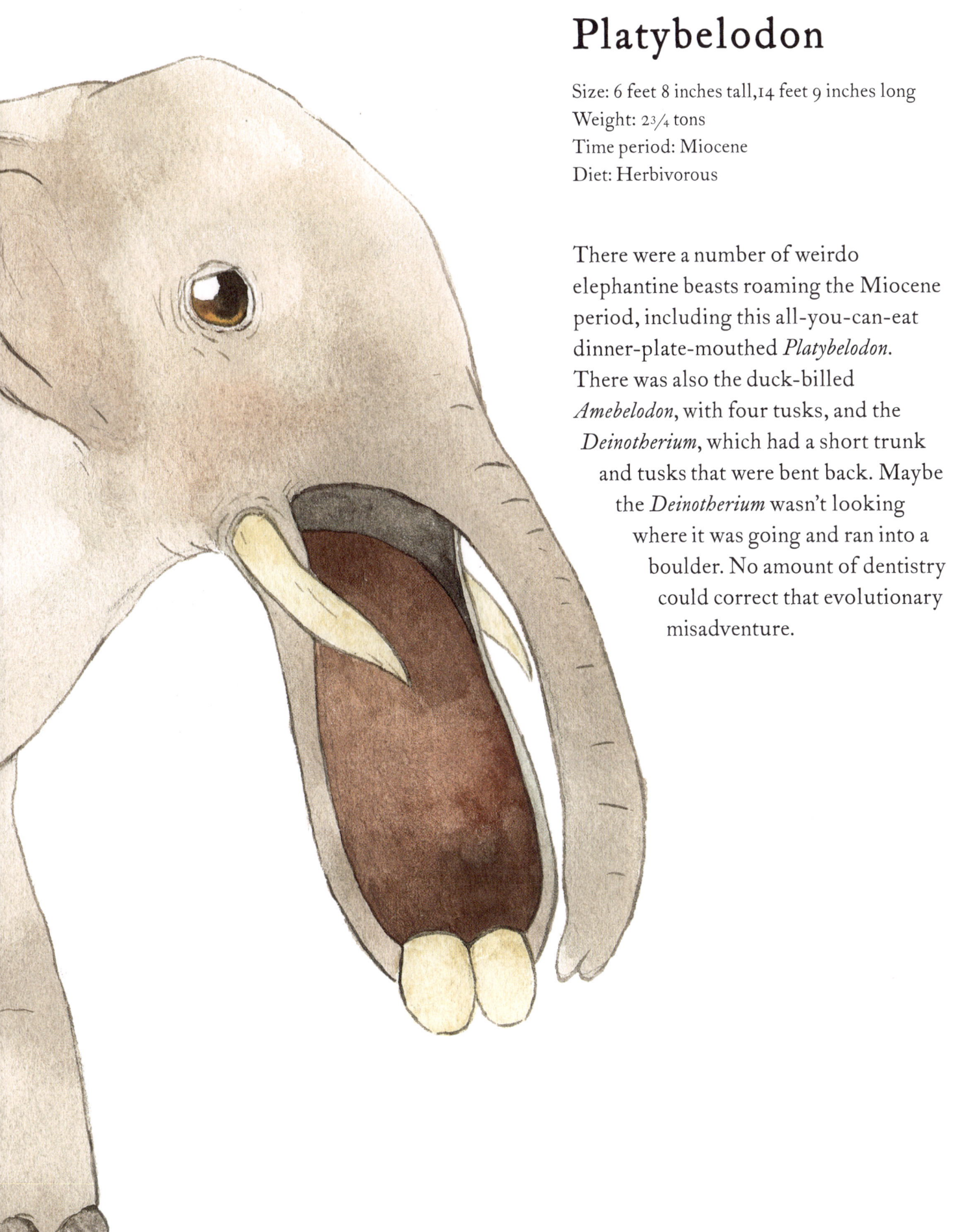

Platybelodon

Size: 6 feet 8 inches tall,14 feet 9 inches long
Weight: 2¾ tons
Time period: Miocene
Diet: Herbivorous

There were a number of weirdo elephantine beasts roaming the Miocene period, including this all-you-can-eat dinner-plate-mouthed *Platybelodon*. There was also the duck-billed *Amebelodon*, with four tusks, and the *Deinotherium*, which had a short trunk and tusks that were bent back. Maybe the *Deinotherium* wasn't looking where it was going and ran into a boulder. No amount of dentistry could correct that evolutionary misadventure.

Bullockornis planei

Size: 8 feet 3 inches tall
Weight: 550 pounds
Time period: Mid-Miocene
Diet: Herbivorous

Not wanting to be left out, Australia had its own fierce squad of gigantic terrifying birds—the largest of which was the 9-foot 9-inch-tall *Dromornis*. But featured here is *Bullockornis*, whose reputation gave it the nickname "the demon duck from hell"! When skeletal remains were found in Bullock Creek, Northern Territory, a deadly looking beak encouraged conclusions that this bird was a predator that would have used its bloodthirsty bill for catching prey and stripping flesh.

We now think that this bird was closely related to wildfowl and would have used its large beak for eating foliage. Vegetarian or not, it is easy to see what the terrified discoverers were thinking when they unearthed and named "the demon duck from hell."

Gigantopithecus

Size: 6 feet to 9 feet 9 inches
Weight: 400 pounds to 1,100 pounds
Time period: Late Miocene to Mid-Pleistocene
Diet: Herbivorous

The great ape—or Mr. Gigantic to you. You would have to be on your best behavior if you met a *Gigantopithecus*, not just because it could be almost 10 feet tall and wide enough to fill a small room, but also because it could be your ancestor. Yes— this big guy is a hominid, just like us. The *Gigantopithecus* lived in Southeast Asia and China until 100,000 years ago, and its favorite food was bamboo—so it probably rubbed shoulders with pandas. I wonder if these giant apes were gentle giants, too, like the peaceful pandas.

Odobenocetops

Size: 6 feet 9 inches long
Weight: 1,430 pounds
Time period: Miocene
Diet: Piscivorous

The *Odobenocetops*, a strange walrus-whale from the Miocene, has a name that translates to "aquatic mammal that seems to walk on its teeth." This sea creature had two tusks, one much longer than the other, and it has been proposed that these were used as primitive echolocations (a sensory system using sound to locate objects) to search for food, because sound would have been magnified into the animal's skull.

Argentavis magnificens

Size: 6 feet 8 inches tall, 20-foot wingspan
Weight: 155 pounds
Time period: Late Miocene
Diet: Carnivorous

One of the biggest birds to have ever flown the skies, the Miocene *Argentavis magnificens* has been linked to the condors of the Americas. Today's Andean condor is an amazing bird with a 10-foot-long wingspan, but the *Argentavis* could have dwarfed it. Its huge wingspan was 20 feet—like a set of black barn doors taking flight. If they had circled in large groups, swirling in great swinging arcs like today's condors, they would have darkened the skies.

Stupendemys

Size: 10 feet long
Weight: About 2½ tons
Time period: Late Miocene to Early Pliocene
Diet: Herbivorous

The leatherback turtle, seafaring and having a teardrop shape, is today's biggest turtle—it measures about 6 feet in length. It is also the fourth biggest reptile in the world. However, the *Stupendemys* was even bigger. This giant freshwater turtle from the Miocene was more than 10 feet long. This colossal South American river dweller would have slowly worked its way through rain-forest streams, grazing on aquatic plants in vitamin-enriched waters. It is hard to believe that a creature so big would have been able to maneuver itself in such tight spaces and not get tangled up in the mangroves and roots. It was around for millions of years, so it must have got pretty good at it.

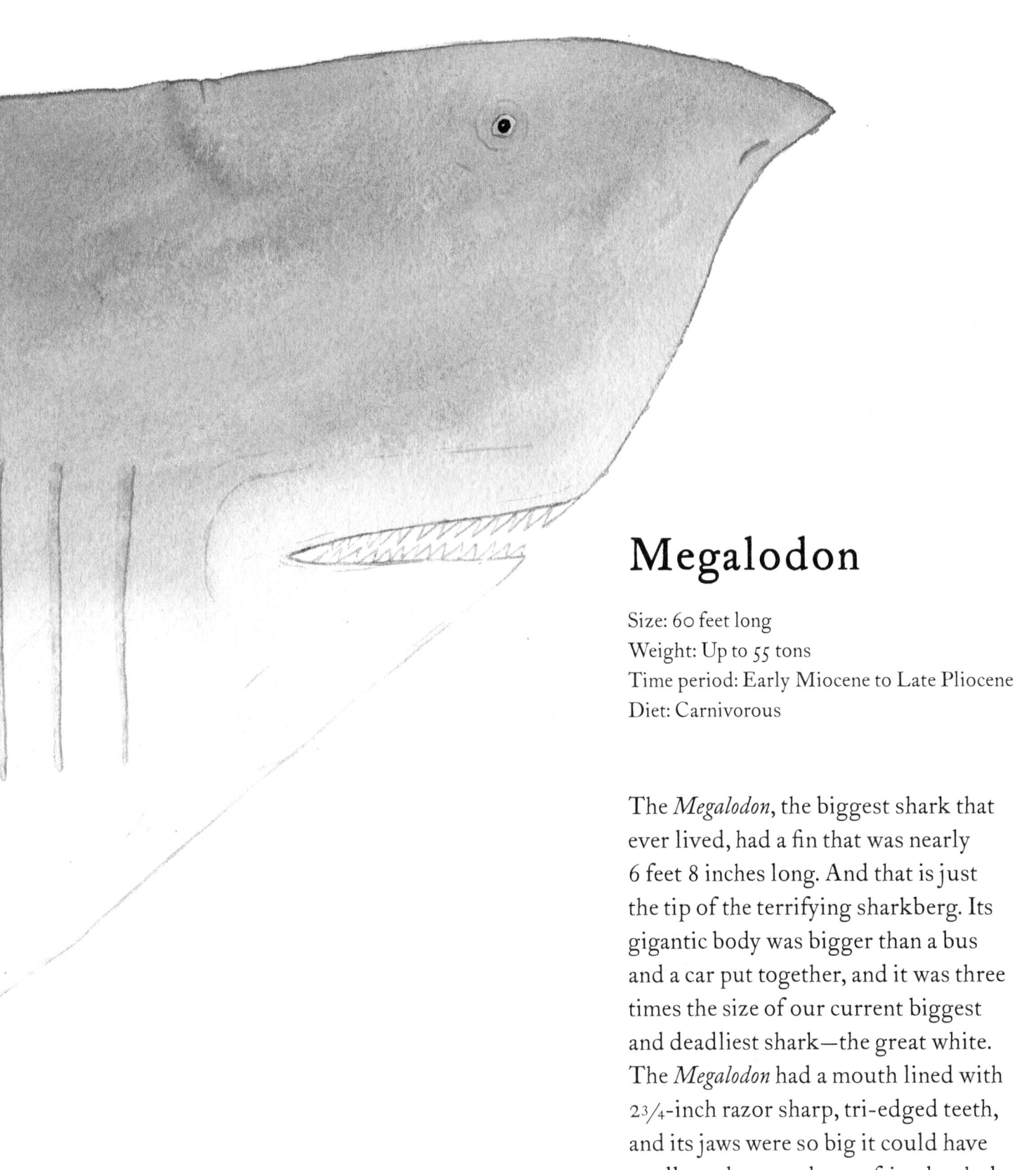

Megalodon

Size: 60 feet long
Weight: Up to 55 tons
Time period: Early Miocene to Late Pliocene
Diet: Carnivorous

The *Megalodon*, the biggest shark that ever lived, had a fin that was nearly 6 feet 8 inches long. And that is just the tip of the terrifying sharkberg. Its gigantic body was bigger than a bus and a car put together, and it was three times the size of our current biggest and deadliest shark—the great white. The *Megalodon* had a mouth lined with 2¾-inch razor sharp, tri-edged teeth, and its jaws were so big it could have swallowed you and your friends whole. Fortunately, all that was millions of years ago.

Megalania

Size: 18 feet long
Weight: 1,325 pounds
Time period: Pleistocene
Diet: Carnivorous

Another exciting member of the Australian megafauna crew, the *Megalania* was the world's largest carnivorous lizard. Some scientists think they were poisonous, too, making them similar to the Komodo dragon, a large toxic lizard of Indonesia today. At almost 18 feet long, the *Megalania* was twice the size of the Komodo dragon, and was big enough to fight a *Diprotodon* (see page 68). It is thought that numbers dropped when humans settled in Australia 60,000 years ago. Komodo dragons are highly dangerous beasts, so *Megalania* must have been a serious threat to human survival. It is clear that either humans or *Megalania* would have had to go.

Procoptodon goliah

Size: 6 feet 8 inches tall
Weight: 440 to 530 pounds
Time period: Pleistocene
Diet: Herbivorous

The *Procoptodon*, or short-faced kangaroo, may
have looked similar to a regular red kangaroo,
and it was actually a hulk. In a regular sitting
stance, these huge roos were more than
6 feet tall and ripped, like an iron-pumping
bodybuilder. They had long front finger claws
for stripping leaves on the hot Australian
plains as well as odd-looking hind-leg claws
ideal for combat and hooflike mobility. Oh,
and a flat face.

Zaglossus hacketti

Size: 3 feet 3 inches long
Weight: 65 to 90 pounds
Time period: Pleistocene
Diet: Insectivorous

Speaking of Australia, they grew them so big there that even their hedgehogs were more than 3 feet long.

Actually, although the *Zaglossus* did have protective spikes and could roll up in a ball, it was not really a hedgehog. This massive echidna was far, far stranger!

The *Zaglossus* had a long beaklike snout and the lengthy tongue and scaly legs of a lizard, just like its much smaller hedgehog-size cousins that still roam the Australian forests and woodlands. It also laid eggs. But the oddest thing about this extinct mammal was that it was the size of a sheep.

Smilodon fatalis

Size: 3 feet 3 inches tall
Weight: 620 pounds
Time period: Pleistocene
Diet: Carnivorous

The Pleistocene poster boy! Probably the most famous of our prehistoric extinct mammals and more commonly known as the saber-toothed tiger (although the *Smilodon* is more commonly related to the lion). Broadly built and powerful, the *Smilodon*'s most outstanding attributes were those astonishing weapons, the saberlike teeth.

Palaeontologists have discovered that the famous fangs were actually fragile and not the weapons of mass destruction you might think. This means that this big cat would have been a patient and precise hunter, killing with stealth and precision instead of overt muscle and force. This lack of blunderbuss power allowed their jaws to be opened freakishly wide, aiding their killing prowess.

Zygolophodon

Size: 13 feet tall
Weight: 15 1/2 tons to 20 tons
Time period: Miocene to Mid-Pleistocene
Diet: Herbivorous

Don't be fooled—this is not a small elephant with long tusks. This is a *Zygolophodon*, a massive elephant with gargantuan tusks. His tusks barely fit on the page. An obvious member of the Elephantidae family, this guy was from the Miocene and Pleistocene periods, just like other famous mammoths that you know. However, the *Zygolophodon* managed to stand out in a crowd, as it had the biggest tusks of all of them (more than 16 feet long). It is not surprising that its name even translates to "long tusks."

Elasmotherium

Size: 15 feet long, 6 feet 8 inches tall
Weight: 3¾ tons to 5 tons
Time period: Late Pliocene to
Late Pleistocene
Diet: Herbivorous

From the extended rhinolike family,
the *Elasmotherium* certainly is not the
biggest—that's the *Paraceratherium* (see
page 36)—but it certainly looks the
meanest. The grown male had a horn
that could have reached a gargantuan
6 feet long, despite being only 6 feet
8 inches tall itself. The *Elasmotherium*
was a herbivore, so this horn was not
used for skewering prey, but instead
for self-defense and showing off to
other males.

In fact, such a big horn has given our
friend the nickname of "The Siberian
Unicorn"—a magical name for a crazy
creature.

(On a side note, please do not confuse
the *Elasmotherium* with the also great-
but-extinct woolly rhinoceros—that
dude is more like a regular rhino but in
a shaggy ginger wig.)

Megatherium

Size: 20 feet long
Weight: 4 1/2 tons
Time period: Early Pliocene to Late Pleistocene
Diet: Herbivorous

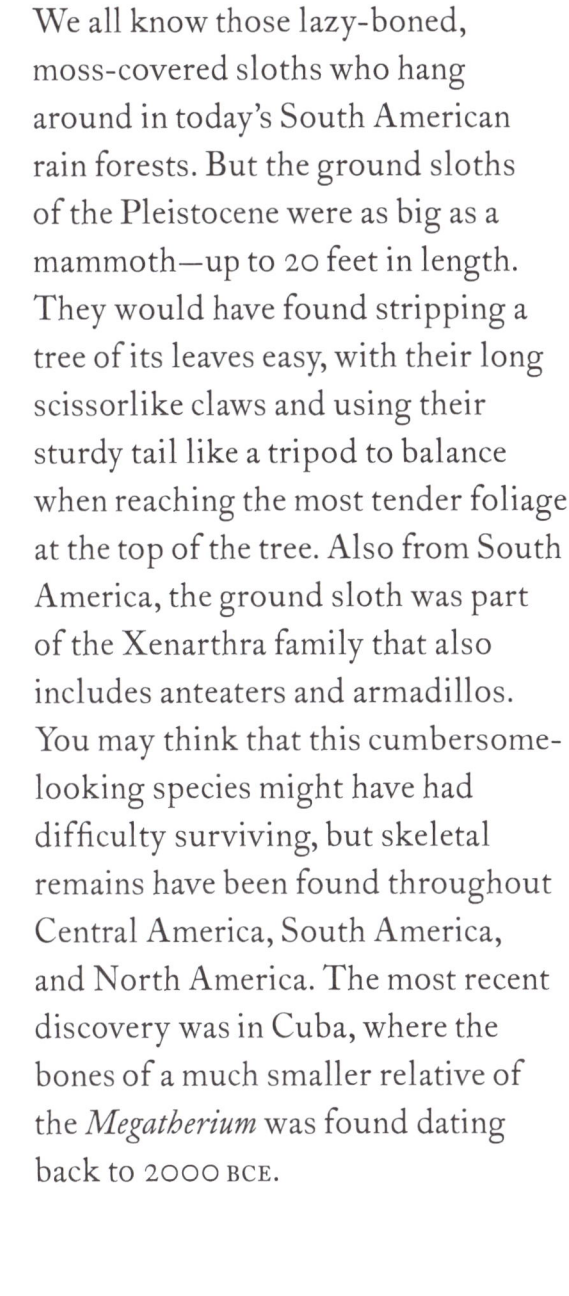

We all know those lazy-boned, moss-covered sloths who hang around in today's South American rain forests. But the ground sloths of the Pleistocene were as big as a mammoth—up to 20 feet in length. They would have found stripping a tree of its leaves easy, with their long scissorlike claws and using their sturdy tail like a tripod to balance when reaching the most tender foliage at the top of the tree. Also from South America, the ground sloth was part of the Xenarthra family that also includes anteaters and armadillos. You may think that this cumbersome-looking species might have had difficulty surviving, but skeletal remains have been found throughout Central America, South America, and North America. The most recent discovery was in Cuba, where the bones of a much smaller relative of the *Megatherium* was found dating back to 2000 BCE.

Diprotodon

Size: 9 feet 9 inches long
Weight: 3 tons
Time period: Pleistocene
Diet: Herbivorous

The *Diprotodon* (meaning "two forward teeth") was the largest marsupial ever known. Just imagine a 3-ton, rhino-size wombat wandering the outback. It was a big beast. Known to have existed throughout most of Australia, the *Diprotodon* died out about 46,000 years ago. What a strange creature it was, with small dainty feet on pillarlike legs, a large nose, and prominent teeth that could have eaten an apple through a thin gap in a fence

Bison latifrons

Size: 14 feet 9 inches long, 8 feet 3 inches tall
Weight: 1 ¼ tons
Time period: Pleistocene
Diet: Herbivorous

An easy indication of the sheer size of the
Pleistocene *Bison latifrons* is to look at the extent of
those staggeringly wide horns. Its descendent, the
modern North American bison, is hardly on the
small size. Those big powerful bovines stand more
than 6 feet high and have a horn span
of 2 feet 2 inches. Compare that to
the *Bison latifrons*, which towered
at 8 feet 3 inches, and had a horn
span of 6 feet 8 inches. It must
have been an awesome sight
to see gigantic herds of these
native buffalo roaming
the plains.

Glyptodon

Size: 13 feet long, 5-foot tall
Weight: 1¼ tons
Time period: Pleistocene
Diet: Herbivorous

Is it a turtle? A dinosaur? A car? At 13 feet long and 5 feet high you could easily be forgiven for mistaking the *Glyptodon* for some kind of new-fangled hybrid automotive vehicle from an alternate dimension. Anatomically, it was similar to an armadillo, but this strange creature was also from the same long evolutionary path as the anteater and *Megatherium* (see page 66). Protected by a fused bony exoskeleton, they would have roamed Pleistocene South America, fighting one another over vegetation patches with their thick heavy tails. Until, like a lot of our megafauna, they mysteriously became extinct at the end of the last Ice Age about 11,700 years ago.

Ornimegalonyx

Size: 3 feet 7 inches tall
Weight: 20 pounds
Time period: Late Pleistocene
Diet: Carnivorous

Like (I hope) everybody, I would have loved to have seen any of the animals featured in this book with my own eyes. But it is the *Ornimegalonyx* that I am most sad about missing. Imagine seeing this giant Cuban owl in its natural habitat, running through the forest and dropping from branches onto unsuspecting prey. This mainly flightless owl stood at more than 3 feet tall and weighed a whopping 20 pounds. In comparison, one of today's biggest owls, the Eurasian eagle-owl, is 2 feet 3 1/2 inches and weighs 6 pounds 10 ounces.

With no need to keep light for flight, the *Ornimegalonyx* used its weight advantage for hunting. Ambush was the preferred tactic—dropping with full force and mighty talons onto rodents, sloths, and even deer. Unfortunately, the introduction of humans and their pets caused the extinction of this mega owl.

Pygmy mammoth

Size: 5 feet 8 inches tall, 14 feet 9 inches long
Weight: 1,675 pounds
Time period: Late Pleistocene to Early Holocene
Diet: Herbivorous

The Pleistocene mammoths of a small island off the coast of California were so isolated that they would have been predator-free. With no threats to their well-being, they did not need size and bulk to protect themselves and so, over generations, they slowly decreased in size. In time, they became their own species—pygmy mammoths. Standing at only 5 feet 8 inches tall, they would have looked small(ish), but they were perfectly adapted to their surroundings. That is, until a catastrophic extinction event wiped out much of North America's mega (and mini) fauna.

Short-faced bear

Size: 10 to 13 feet tall (on hind legs)
Weight: 2,200 pounds
Time period: Mid-Pleistocene to Early Holocene
Diet: Omnivorous

What a monster! I would not have named
this beast after its short snout when there are
so many other ferocious features to take into
consideration. The short-faced bear was huge,
had long muscly forearms (with a 13-foot reach)
and long agile hind legs. This meant that the bear
could grapple and grab prey in the blink of an
eye, but it could also sprint and was fast enough
to catch an antelope on the hoof. It is thought that
the short-faced bear is the reason why humans
were late arriving in North America, because
these animals lived in the Bering Strait, patrolling
for food. For early man, this small landmass was
the only entrance to North America, because
it connected it to the rest of the world via
Russia, and these bears would have been
fearsome guards.

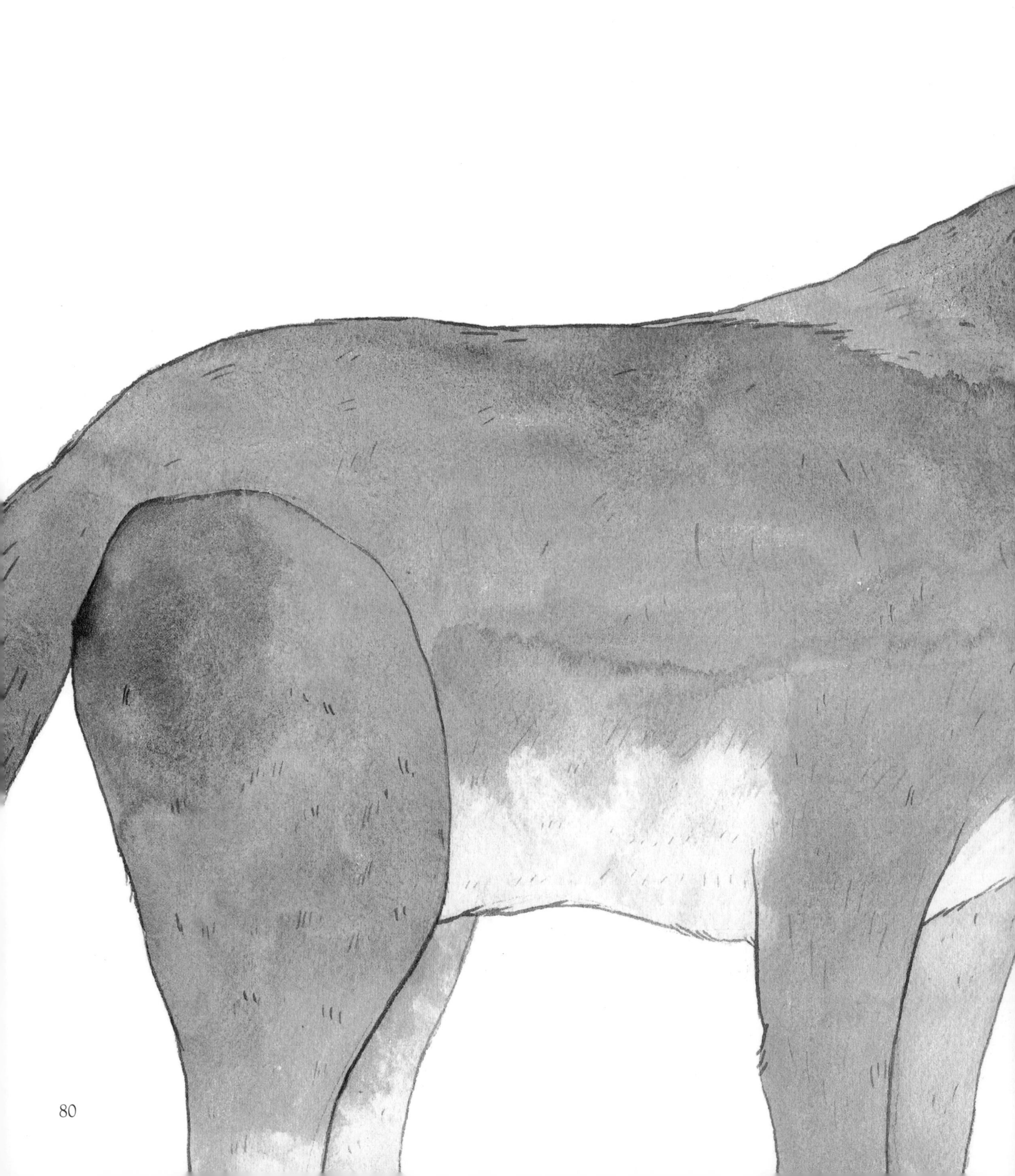

Dire wolf

Size: 5 feet long, 3-foot tall
Weight: 130 to 175 pounds
Time period: Late Pleistocene
Diet: Carnivorous

The Canidae family is thought to originate from North America. This includes all wolves and dogs—jackals, dingoes, coyotes, labradors, border terriers, and even pugs. The dire wolf here was the fearsome granddaddy of all of them. Built only a little bigger than today's common gray wolves, this lupine legend had an incredibly powerful bite for its size. Working cooperatively in large packs, the wolf could easily hunt huge prey of the Pleistocene, such as mammoths and *Megatherium*. Not bad at all, but could it fetch a stick?

Sivatherium

Size: 16 feet long, 10 feet tall
Weight: Up to 1⅓ tons
Time period: Pliocene to Holocene
Diet: Herbivorous

The *Sivatherium*, the largest member of the giraffe family, stood nearly 10 feet tall, not counting its huge crown of horns and strange forehead protrusions. This mooselike plant eater's name means "Shiva's beast," from one of the three creation gods from the Hindu faith. Although thought to have died out a million years ago, an ancient Indian civilization may have known this beast in person. Cave paintings have been found appearing to depict the distinctive *Sivatherium*, most recently from Sumer— one of humanity's earliest civilisations. The 5000-year-old Sumerian Kish statue is shaped like a stag, and he looks really similar to our beast here. It is interesting to think how many of our forgotten beasts met our early ancestors in person.

Megaloceros

Size: 11 feet 8 inches long, 6 feet 9 inches tall
Weight: 1,200 pounds to 1,325 pounds
Time period: Mid-Pleistocene to Early Holocene
Diet: Herbivorous

Commonly known as the "Irish elk," the *Megaloceros* was neither from Ireland nor an elk. In fact, this huge deer was once found everywhere east of Ireland—from the United Kingdom to North Africa and even in China. The male *Megaloceros* was more than 6 feet tall and had antlers that were fitting for the largest deer ever to have lived. These monstrously large adornments were works of natural art—breathtaking, powerful, and staggeringly big. Some measured 12 feet wide. There are many theories that attempt to explain the extinction of the Irish elk, including one that hypothesizes that the antlers got so big that the animals could not walk, nor mate with one another. The most reasonable theory, and one that we can particularly relate to, is that climate change and human activity were to blame.

Woolly mammoth

Size: 10 feet tall
Weight: 6⅔ tons
Time period: Pleistocene to
Early Holocene
Diet: Herbivorous

The mammoths were a strand of the large umbrella family of Elephantidae that has existed for more than 5 million years (and includes today's elephants). Around the same size as an African elephant, the woolly is instantly recognizable by its shaggy, dark orange fur coat and tusks like curly lances, which look protective and deadly at the same time.

They were much smaller than the steppe and southern mammoths who were alive at a similar time, but the woolly was so well equipped for the ice age that its range took in most of the northern hemisphere, including North America, the United Kingdom, Europe, Russia, Siberia, and China.

They existed at the same time as early man, who hunted them for their bones, pelt, meat, and tusks. But it was climate change and territory and habitat shrinkage that dispersed the woolly mammoth and decimated its population. They were pushed to small island outposts and actually survived as recently as 4,000 years ago.

Meiolania

Size: 8 feet 3 inches long
Weight: About 1,000 pounds
Time period: Mid-Miocene to Holocene
Diet: Herbivorous

This spotted fellow hails from the kingdom of New Caledonia and Eastern Australia from the bygone times of the Miocene. A turtle armored to the teeth, maybe a little too exuberantly so? Those helmet horns would have prevented our hero from withdrawing its head into its shell, which is a famous turtle self-defense tactic. Maybe this played a part in its downfall?

Japanese Honshu wolf

Size: 3 feet long, 1-foot tall
Weight: 40 pounds
Time period: Holocene
Diet: Carnivorous

In Japanese folkore, the Honshu wolf is the guardian and protector of travelers. It was also one of the smallest wolves ever—measuring only 3 feet from nose to bushy tail. A distant cousin of the North American gray wolf, this beast eventually settled on a group of small Japanese islands. The Honshu thrived until the 1700s, when they were wiped out by rabies—a human import. However, like many of our wonderful and sadly extinct beasts, there have been reports of sightings—and even possible hidden enclaves. Let's hope they have not been doomed to folklore just yet.

Elephant bird

Size: 10 feet tall
Weight: 880 pounds
Time period: Holocene
Diet: Herbivorous

The elephant bird (*Aepyornis maximus*) was an enormous flightless bird from the island of Madagascar. It had legs like tree trunks, stood more than 10 feet tall, and laid eggs as big as footballs—160 times bigger than a chicken egg. It was a true giant of the bird world, perhaps for millions of years, until humans inhabited its island and it was slowly wiped out only about 1,000 years ago.

South Island giant moa

Size: 6 feet 8 inches to 10 feet tall
Weight: 550 pounds
Time period: Late Pleistocene to Holocene
Diet: Herbivorous

The amazing moa is a family of nine species of flightless, wingless birds from New Zealand. The South Island giant moa (*Dinornis robustus*) was the biggest of these. It was 6 feet 8 inches tall and extremely heavy—about as much as a huge, fat pig. Unfortunately, the moa went the same way as many pigs—they were all eaten within a hundred years of humans arriving on the islands. The only moa remaining after the 1300s were those in Maori folk tales.

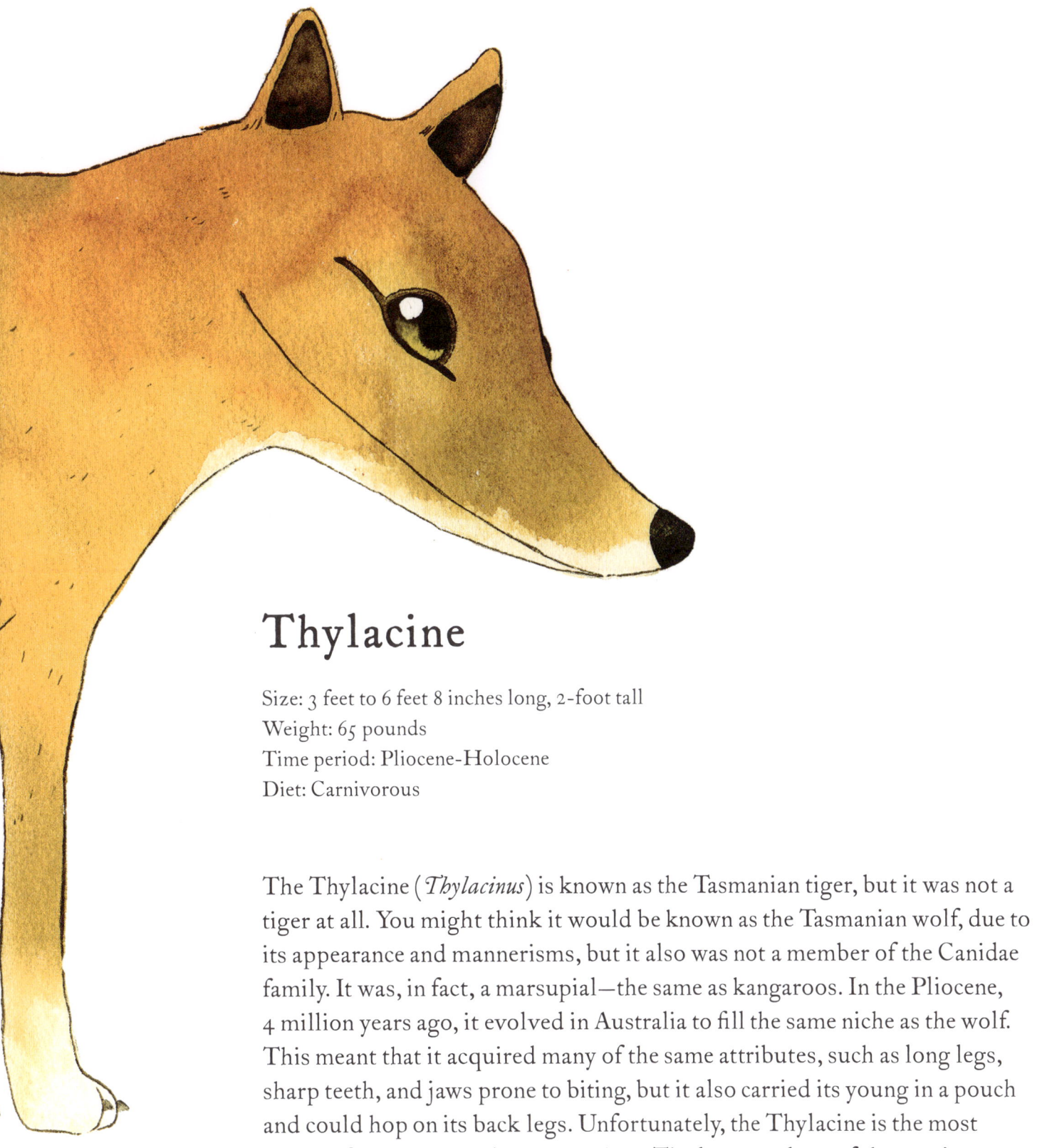

Thylacine

Size: 3 feet to 6 feet 8 inches long, 2-foot tall
Weight: 65 pounds
Time period: Pliocene-Holocene
Diet: Carnivorous

The Thylacine (*Thylacinus*) is known as the Tasmanian tiger, but it was not a tiger at all. You might think it would be known as the Tasmanian wolf, due to its appearance and mannerisms, but it also was not a member of the Canidae family. It was, in fact, a marsupial—the same as kangaroos. In the Pliocene, 4 million years ago, it evolved in Australia to fill the same niche as the wolf. This meant that it acquired many of the same attributes, such as long legs, sharp teeth, and jaws prone to biting, but it also carried its young in a pouch and could hop on its back legs. Unfortunately, the Thylacine is the most recent of our beasts to become extinct. The last members of the species were kept in Hobart Zoo, Tasmania, until 1936, from where we have photographs and film footage of the dearly departed. These serve as stark warnings of the consequences of humankind's growth and expansion across the earth.

Thank yous

To Jess, for her endless love and support, and Romy, Mae, and Bonny, for being the best ever.

To the Sewells, the Roses, O'Sullivan-Averys, and the Lees.

To Dr. Stephen Brusatte of The University of Edinburgh, Matthew Bragg, Pablo and family,
the Okada-Hoods, the Bucklands, Jo at The Hayloft,
Ray Harryhausen, Dougal Dixon, Graham Hancock, and John Anthony West (RIP).

PAVILION

First published in the United Kingdom in 2018 by
Pavilion Children's Books
43 Great Ormond Street
London
WC1N 3HZ

An imprint of Pavilion Books Limited.

Publisher and Editor: Neil Dunnicliffe
Assistant Editor: Harriet Grylls
Art Director: Anna Lubecka
Scientific Advisor: Dr. Stephen Brusatte

ISBN: 9781843653936

10 9 8 7 6 5 4 3 2

Reproduction by Mission Productions Ltd., Hong Kong

Printed by Toppan Leefung Ltd., China